# Einstein's Special Relativity: Discover the hidden secret of Time.

# How to slow the speed of Time.

**Wali Charles PhD**

**Copyright © 2019 WALI CHARLES PhD**

All rights reserved. No portion of this book may be reproduced in any form, stored in any retrieval system, or transmitted in any form by any means—electronic, mechanical, photocopy, recording, or otherwise—without prior written permission of the publisher, except as provided by United States of America copyright law.

## Dedication

To the giants on whose shoulders' I'm standing.

- Sir Isaac Newton

# CONTENTS

## CHAPTER ONE    1

Einstein's Discovery

Newton's Three laws of motion

Where Newton got it wrong

## CHAPTER TWO    8

The long journey to 1905

Gedankenexperiment: Einstein's Day Dreaming

Einstein's first theory: On the Electrodynamics of Moving Bodies

## CHAPTER THREE    18

Einstein's first theory: The Consequences

Length Contraction

Time Dilation

Twin Paradox: A proof of Time Dilation

What other proof?

## CHAPTER FOUR    31

Speed of Light: Can we achieve it?

Speed of Light: Einstein's Thought

## CHAPTER FIVE 36

What Next?

# CHAPTER 1

Yes, you read the title right; time has a hidden secret. I walk around every day, and it baffles me how much people know nothing about **Time**. Yet, it's secret has been lying in books on our shelves for more than 100 years now. Have you never imagined how sci-fi movies came about the idea of traveling through time? Or the logical explanation behind the popular Marvel's series – FLASH. How could anyone think of a man being able to move at such a speed?

I'm going to reveal to you the secret of Time. How about if I tell you, light can be bent? That's impossible; I'm sure that's what you will say. I won't be talking about the art of bending light in this book, though. Watch out for my next book. I will be releasing books on these fantastic possibilities as time goes by.

I'm sure at the end of this short book; you will be amazed. Then, you will understand why Albert Einstein is well respected in the science world.

Don't worry, we won't go into much of the technicalities, but just enough to get you informed. If you need more information, you know what to do – hit the library, get the bigger guns (physics textbooks).

There, you will find lots of theoretical proofs (mathematical equations) to play around with. Here, I will be talking more about the physical proofs and other awesome physics you won't believe. Note, these are all real-life practical phenomenon, not magic or fantasy.

## 1.0 Einstein's Discovery

First, we must trace back the history of one of the most mind-blowing discoveries in the history of science – **Relativity**. Ever heard of the scientist called Albert Einstein? I'm sure you have. He is the man with the crazy hair and tongue out, you see in a famous picture he took on 14th Of March 1951, while leaving his 72nd birthday party. The photo was taken by one of the photographers hounding him around, Arthur Sasse. Unfortunately, that picture is copyrighted, so it can't be included in this book.

The concept of time, space and how light travel as we know today, are based on his work, and that of many other scientists. But his' was quite revolutionary for his time, because there was no experimental physical proof as at then. It was all theoretical physics used to explain observed natural events, which couldn't be duplicated in a lab.

## 1.1 Newton's Three laws of motion

Before Einstein's special relativity theory of 1905, where he explained the relationship between space and time, for a body moving at the constant speed of light ($3 \times 10^8 \, m/s$) – as earlier discovered by James Maxwell in

1865. Scientists understood the laws of nature based on the three laws of motion, as postulated by Isaac Newton in 1689.

i. A body at rest or in motion will remain in this position unless an external force is applied on it. This is generally called the laws of inertia.

ii. Force is equal to the product of mass and acceleration. i.e., The force acting on a body will produce an acceleration which is directly proportional to the "amount" of the force, but inversely proportional to the mass of the body, which means the heavier the body, the more force required to move it.

iii. For every action, there is an equal but opposite reaction.

A corollary from these laws was the law of gravity also discussed by Newton. He theorized that gravity is an inherent property of bodies. For instance, the gravity from the center part of the earth is the force keeping us from floating away from its surface. This same force is what causes the famous apple to fall from the tree. Also, he concluded that this force called gravity is what keeps the earth "locked" in its revolutionary path with the sun. That is, every object has a gravitational force, with which it attracts other bodies around it. The more massive the body is, the stronger it's attractive force. Although the less massive body is exerting some force also, that of the more massive body will be the determinant of the resulting force from the interaction. Whoops! Sounds tricky? You can read it again.

Conclusively, Newton from his work was saying; time is absolute - it can't be acted upon by any force – it flows equably. Then for space, he said, it is immovable and similar everywhere. These were the laws that explained everything about the physical world as we knew it then.

But, as much as these Newton's laws were revolutionary for his time, scientists were noticing some divergence from these laws in some of their experimentations in the lab. As you can guess, scientists were puzzled. But was Newton, right? Keep reading.

Not to take away from Sir Newton's contribution to science, his laws held their own in the following instances:

- For objects moving at speed very low compared to the speed of life.
- For much larger objects compared to sub-atomic.
- For less dense objects, i.e., the famous falling apple, human beings, a truck – just everyday life bodies, but not black bodies, or stars.

All these three characteristics must be taken into account before Newton's laws can be applied to a problem successfully.

## 1.2 Where Newton got it wrong

Newton never was puzzled by the source of gravity, which he talked about as a part of his laws. Personally, I feel, had Sir Newton considered the works of the Scottish philosopher David Homes, A Treatise of Human Nature, he would probably have not made the error he did. By the way, you can read Homes' work online. I think you should. It's quite interesting for the curious mind. This same work was what Einstein referred to in a letter recently discovered at University of Edinburgh, around February 2019, "It is very possible that, without this philosophical study, I would not have arrived at the solution"

Homes believed that scientific concepts shouldn't be based on reason alone. They should be based on experience and evidence. This was exactly what Einstein did in 1905. Isn't it amazing how a book written 166 years before 1905 influenced the face of science?

Back to our focus, where did newton get it wrong? You must keep it in mind that, before 1905, the speed of light had been measured. During this time, most scientists believed light needed a medium for propagation. It's only logical to expect this since experiments had concluded that sound couldn't travel in a vacuum. You can confirm this yourself with a very simple experiment. Get a vacuum (i.e., a bottle sealed tight, with no air in it) with a bell in it. You will observe that, no matter how much the bell jingles, no sound from the bell will be heard. Proving sounds require a medium to propagate since it propagates through air, water, and solid substances. Elegant reasoning!

Scientists then concluded this same concept must apply to light also. Since light travels through vacuum, then there must be a medium in the vacuum to allow such. Fantastic line of reasoning! Don't you think?

# CHAPTER 2

## 2.0 The long journey to 1905

From the 1800s to the early 20th century, the drive of the physics world before 1905 could be said to have been largely based on proving the existence of the elusive **Ether or Luminiferous Ether**. The medium believed to be responsible for the propagation of light through space and vacuum. Experiments were being designed and carried out to confirm the existence of this invincible medium. There was also a very significant development about the duality of light as an electromagnetic wave, but we won't go into that here. Kindly bookmark my page to read about it soon. Be assured; it will blow your mind.

Since Earth is in constant motion around its orbit and on a revolution around the Sun, two major theories about how Ether interacts with nature were considered by scientists.

1. Ether is stationary and only partially dragged by the earth due to its (earth's) motion. Augustin-Jean Fresnel proposed this in 1818. He is the inventor of the Fresnel lens, used in lighthouses to extend the reach of the light, until the mid-20$^{th}$ century. The lens is still used in some cars reverse light, left-hand-driven trucks entering Europe from the UK and airplane's cabin light system.
2. Ether is dragged by Earth completely. Meaning, it's like an invincible water in a bottle. The movement of the bottle entirely determines its movement. This was proposed by Sir George Stokes in 1844.

With the Fizeau experiment by Hippolyte Fizeau in 1851, to measure the relative speed of light in moving water. The 2$^{nd}$ theory was proved false. The thought experiment behind the Fizeau's lab work was a simple one. Since light has a speed in any given medium A. When it is propagated through the moving medium A (medium A standing for a completely mobile Ether), the speed of the light should be a summation of its (light's) speed in that medium A, and the speec of the moving medium A. Very straight forward, and simple. The experiment proved the

2nd theory to be wrong. From the experiment, it was found out that the speed of the light through the medium A, didn't change based on the movement of the medium. Hence, the presence of a completely dragged Ether, influencing the propagation of light, couldn't be proved.

Following this experiment, the 1$^{st}$ theory was stuck with through the years between 1851 to 1905. Within these years, other experiments were conducted to verify the existence of a partial inertia ether. The most famous was the Michelson–Morley experiment of 1887 by Albert A. Michelson and Edward W. Morley. It was another simple experiment involving measuring light's speed traveling at different angles. The logic behind this was since ether is partially stable, it shouldn't be equally distributed. Meaning, there should be some noticeable difference in speed of the light, when light is made to travel through different angles and paths. Sadly, this experiment and others, all came out with no proof of the existence of an invincible partially stationary medium that propagates light.

I believe it should be noted that these were all brilliant scientists. Their works were contributing to the body of knowledge of physics, as it was known then. Albert A. Michelson is the first American to win the Nobel Prize in a science field (1907). He was the founder and the first head of the physics department of the University of Chicago. His colleague, Edward W. Morley, was also an American scientist. He is famous for his exact and accurate measurement of the atomic weight of oxygen using the wet chemistry method, widely acknowledge has been laborious.

## 2.1 Gedankenexperiment: Einstein's Day Dreaming

In comes Einstein in 1905 with a solution to explain this problem. Einstein did what the other scientist wouldn't do. He accepted the results from the experiment as they were while trying to deduce a theory based on the results they produced. This was the turning point for science.

Just like the Scottish philosopher David Holmes advised in his book Einstein had read – scientific concept should be based on evidence and experience, not reason. All he had to do was come up with a thought experiment verifying the reality of how light behaved, then prove this theoretically with the mathematical concepts that were available; at least those concepts that agreed with the experimentation observations reported. The result was a new understanding of the laws of nature since the laws of physics will always hold everywhere.

He would later claim in 1968 that as a 16 years old boy, he had a thought experiment, which had always fascinated him. This thought experiment assumed he is moving at the same speed with a light beam, and when he looks at a parallel beam, the beam should appear frozen since they are of the same speed. This means he has always been curious about the behavior of light. Einstein is famous for using this sort of thought experiment called **Gedankenexperiment** in German, to solve scientific problems. The compound word "thought experiment" is a calque of the German word.

The thought experiment can be a little hard to imagine, I know. How about you close your eyes and imagine yourself in a moving bus/car. Now open your eyes and look to the floor. Imagine the car's window is shut and tinted; you can't see outside the car. Also, imagine the car's engine is silent, and there's no sound at all. Does the floor of the car appear moving to you? Of course not. You know why? It's simply because you are moving at the same speed as the car. This is what he was talking about.

But this wasn't the case from all the experiments conducted in his time as a teenager – as it especially didn't conform with Maxwell equation of electromagnetism through an ether. Maxwell's laws were the alpha and omega of everything that had to do with electricity, magnetism, and light as known in physics in 1895. The laws were strict, which made teenage Einstein feel there were a lot of questions unanswered. Like why there can't be stationary fields(light), even when an observer achieves the speed of light.

This meant to him that there was an erroneous assumption in the laws of physics as at then. This line of thought would continue to haunt him, all through his university days at Eidgenössische Technische Hochschule Zürich (ETH), and his move to Bern, as a Swiss patent examiner.

I'm sure you can see now that the leaps in science aren't made in one day. It is always a result of lengthy-time of brainstorming and relying on earlier works.

## 2.1 Einstein's first theory: On the Electrodynamics of Moving Bodies

In his published paper of 1905 titled "On the Electrodynamics of Moving Bodies," popularly known as **Special Relativity Theory**, Einstein made two postulates for his theory.

i. There is no uniform frame of reference. The concept of motion is relative, as there is no one fixed reference point. This means the idea of absolute rest and the ether have no meaning. Even a body at rest, is also on motion, since the earth is orbiting around its axis, and on revolution around the sun.
ii. Light is propagated in space with a speed which is independent of the motion of the source. The speed

of light from a fast-moving bulb will be the same as that from a stationary torch.

With these postulates, Einstein was able to explain why the speed of light in a moving frame of reference, like a train, would not be different from that not on the land, as measured by an observer on the train and land, respectively. Remember, experiments measuring the speed of light had been done on the moving frame of reference, which found no difference in light's speed. Many scientists had done experiments in different mediums, which should naturally influence the speed of light, but it never happened. Also, the concept of an invincible medium influencing light speed was famously put to rest by the Michelson–Morley experiment of 1887. All these earlier works provided the perfect platform for Einstein to come up with his theory of relativity.

Einstein had a breakthrough in thought in May of 1905 while discussing with his friend Michele Besso. He would later acknowledge his friend at the end of his would-be famous paper "I am indebted to him for several valuable suggestions"  In the paper, Einstein used this same

**Gedankenexperiment** below, to explain Special Relativity:

"If lightning bolts were to strike both ends of a train, at the same time. An observer at the embarkment will see the resultant light from the two events as simultaneous. That's if the observer is located at the midpoint of the train. This is because the light from the strikes will travel the same distance to reach the observer's eyes. And since they (the light from the lightning strike) are of the same speed, they will arrive at the same time. If another observer were to be on the train, seated at the exact midpoint, he would also measure the speed of light to be the same, as confirmed from earlier experiments on trains. But, since the train is moving forward, the light from the strike at the back-end of the train will have to travel a longer distance before arriving at the observer's eye, compared to the one from the front-end of the train. Why? Because the train is moving forward. It's more or less like the train is trying to "run into the arms" of the light traveling from the front-end of the train. This will make the light from the front-end, arrive at the eyes of the observer first, before that from the back-end of the train.

"

This was the thought experiment that confirmed to Einstein that time is relative since simultaneity is relative. An event can coincide in a reference frame, while in a second reference frame, it would be two separate events. Meaning, time isn't absolute, as assumed by Newton. As can be seen from the experiment, the observer on the embarkment experienced the lightning strikes "correctly" as being simultaneous, while the observer on the train, experienced the lighten at the front-end as occurring first, before the one at the train's other end.

This was all he had to prove his theory, aside from the exceptional mathematical equations, of course. His **Gedankenexperiment** might take some time to understand; you should probably reread it, and try to close your eyes to visualize it. Is it quite amazing?

# CHAPTER 3

## 3.0 Einstein's first theory: The Consequences

In theoretical physics, once equations are derived to explain a phenomenon, there are always consequences that can be obtained from those equations even when they've not been seen or experienced.

## Length Contraction

The first consequence of the theory is called **Length contraction.** It merely states that, when a body moves close to the speed of light, it's length contracts in the direction of its motion. I wouldn't talk much about this, because due to the limitation of our technological advancements right now, this phenomenon can't be measured directly. Now, imagine a witch on a broomstick, oh yes she's on her way to Hogwarts School of Witchcraft and Wizardry. If she's moving at the speed of light or close to it (what we call relativistic speed), from Einstein's theoretical mathematics, it will be observed that both she and the broomstick will shrink along their path of motion. Meaning, if the stick were 20cm before, now it would be 5cm. So far, this hasn't be confirmed directly.

Hopefully, sometime in the future, it would be possible to be able to measure such a body. It will be useful to keep in mind that, the witch and the broomstick will appear contracted (shrink) only when viewed by a stationary observer, relative to her motion. So, the witch herself with the relativistic speed, won't feel any different.

## Time Dilation

This is the second consequence of the Special relativity theory. Imagine two astronauts, Mike and Allen. They both are traveling in different spaceships; yes, they are Star Wars captains, haha. Astronaut Mike is bored. He decides to play with his laser light from his keyholder. He points the red leaser light to a mirror. A beam of light immediately leaves the leaser and hits the mirror. As expected, the mirror reflects the light to the floor. Let's assume it takes 1 second for this to happen. We both agree nothing extraordinary is happening here.

Ok, Astronaut Allen is bored also. He decides to look out of his window. His spaceship is faulty, so it's moving on a lesser speed compared to the other spaceship. Just when he turns to the window to feed his eyes, Mike's spaceship speeds past him. This is where the twist comes in. According to Einstein's equation from his explanation of relativity, if astronaut Allen could see through his window into the spaceship of Mike. Allen will observe something a lot different from what Mike is observing about the leaser beam's movement.

Astronaut Allen will observe the leaser beam travel at a much greater angle towards the mirror, then to the floor. Based on this observation, Allen would have experienced a longer path as being followed by the beam. Let's imagine there was a timer on both spaceships. The measured time for the beam to hit the floor, as observed by Mike and Allen, would be different also. Astronaut Allen's timer would read a higher time (say 2 seconds instead of 1 second). Does this mean time slows down in Mike's spaceship (reference space), as seen from Allen's spaceship? Or what does this mean? Wait for it.

Remember, the speed of light is constant in space. Then, there's no way you can say the light beam in Mike's spaceship is at a different speed when seen from Allen's spaceship. How then would you give meaning to this natural phenomenon? This is the consequence of the theoretical equations worked out by Einstein.

It is real. The mathematics adds up, and this theory by Einstein has been proven over and over again, over the years, even in nature itself. The only possible explanation is that when a body, i.e., Mike's spaceship, is moving at or close to the speed of light, and another body, i.e., Allen's spaceship, is of lower speed or stationary in relative to it, then the two ships will experience time differently. Meaning they are in two different reference frames relative to each other. This implies time slows down, the faster you go, when viewed from a much slower or stationary reference frame. Oh, my goodness, how can this be! You may say, it's not logical at all, I know this also.

You need to understand that Mike would experience the beam has been "normally" behaved. And, if Allen were to have the same beam in his spaceship, he would experience it has been "normally" behaved also. The only reason a divergence from normal (an abnormal) observation was made is because of the speed difference between the two spaceships.

It's crucial to emphasize here, this only occurs at the speed of light, or close to it. One of the spaceships has this speed, while the other spaceship is of a lower speed. The more difference in speed between the spaceships, the more the extent of abnormality observed. i.e., higher time taken for the beam to complete its journey as viewed from the other spaceship. The angle of reflection will be much more different certainly. Which in effect means, the slower the time in the spaceship with the faster speed.

Ultimately, this would mean, if our bored astronaut Mike were to be the one peeing into the window of the astronaut Allen's spaceship, he would observe something different also. If Allen were to be bouncing off a beam on a mirror too, it would take much less time for the beam to be reflected to the floor. This means time would be moving faster in Allen's spaceship when viewed from Mike's spaceship.

This phenomenon is called **Time Dilation**. It happens in our everyday life, but it becomes noticeable at speeds close to the speed of light. Check this illustration.

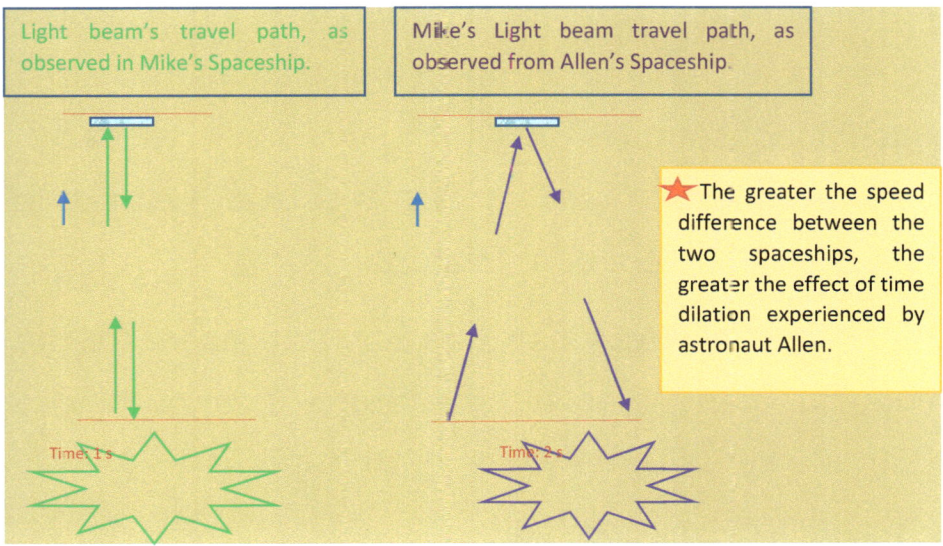

## 3.1 Twin Paradox: A proof of Time Dilation

Oh, yes, it can be proved. Happy? Time dilation wads confirmed by a joint Russian and National Aeronautics and Space Administration (NASA) one-year mission to the International Space Station (ISS). It took place between 2015 – 2016, with one astronaut from each of the US and Russia. At the end of the voyage, after spending a total of 520 days in space (340 days on the

ISS) and returning to Earth on March 2, 2016, the American astronaut, Scott Kelly, was found to have aged slower compared to his twin brother, Mark. Amazing! Right? The protective caps (Telomeres) of some of Scott's chromosomes were found to have elongated as measured from his blood samples when he got back to earth. Generally, telomeres should shorten with age, just like in Mark's. But Scott's were a lot longer.

This was all part of NASA's Twin Study, to prepare for the expedition to Mars, in no distant future. The journey to Mars will take about three years, earth time.

Just in case you are in doubt of findings, both brothers were astronauts, so they mostly had the same life routine, before the voyage. Also, tests were performed on both, to know their health state before Scott left for the ISS. And even during the mission, blood and tissue samples were being taken from both brothers to monitor their metabolism and cell growth. Too bad the Russian cosmonaut, Mikhail Kornienko, doesn't have a twin. It would have revalidated the time dilation phenomenon.

The extent of the difference in age wasn't much, just as predicted by Einstein's theory. Remember that the speed of light must be achieved with respect to a relatively stationary body for a much noticeable difference to be observed. Sadly, while in space, Scott was only moving at a fraction of the speed of light, zooming around Earth at 17,500 mph (28,160 km/h). If only he could have been moving at the speed of light, what an interesting scenario that would have played out!

The answer to living a lot longer is simple then. Technology will have to catch up with the discovery of being able to maintain living beings on a continuous path of relative motion close to the speed of light.

## 3.2 What other proof?

Do you need more proof? Another proof of time dilation is encountered in our everyday lives, especially when using our phones and cars. It's called popularly referred to as the Global Positioning System (GPS). There are three major systems in operation globally right now. The American system of satellites called the NAVSTAR GPS, the European Union's Galileo satellite systems, and Russian GLONASS. The Chinese will be expanding their (BeiDou-2 system) to a global scale in the early 2020s.

How does GPS prove time dilation? First, we need to know how GPS works. The standard satellite GPS configuration is made up of 24 satellites orbiting around the Earth. The satellites are at an altitude of 20,000 km from the surface of the earth, moving around the earth with a speed of 14,000 km/hour.

These satellites are distributed to have a minimum of 4 (max. of 12) of them being visible from any point of the Earth, at any given instance.

Each satellite has in it, an atomic clock. This is the most accurate timepiece in the world. It has an accuracy of 1 nanosecond, i.e., 1 billionth of a second. Not that we have much interest in all these details, but the concept is straightforward. The GPS receiver in the tractor below, or phone, or anywhere on Earth, can know it's position by comparing the time signal it receives from the atomic clocks in each of the satellites in space.

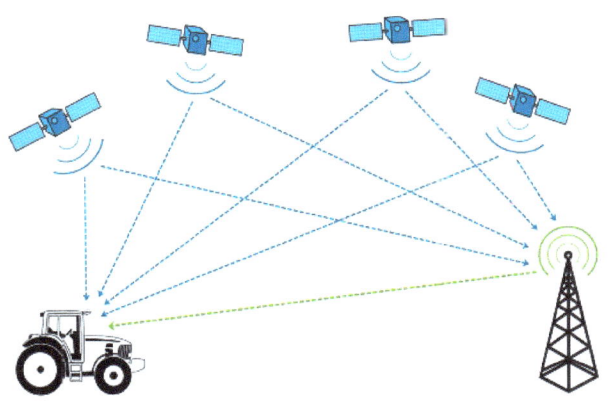

The more the satellites directly above the GPS receiver on the Earth's surface, the more precise the location/position of the GPS receiver can be revealed. When the first satellite broadcasts its signal, the GPS receiver detects it. The receiver syncs its timing with that of the incoming signal from the atomic clock, to calculate it's (GPS receiver) distance from it. This isn't enough information, though. The receiver could be anywhere within the circumference of the broadcast of the signal. Why? All distances within the circumference of the broadcast to the center of the circle are equal.

Then, the second satellite does the same thing, and the possible area where the receiver can be is reduced; It could be any of the two points where the two circles of the broadcast intersect. This continues for all the four satellites. The location where all the satellite intersects pinpoint the specific area where the receiver is. The more satellites overheard the receiver, the better. You can see the illustration below.

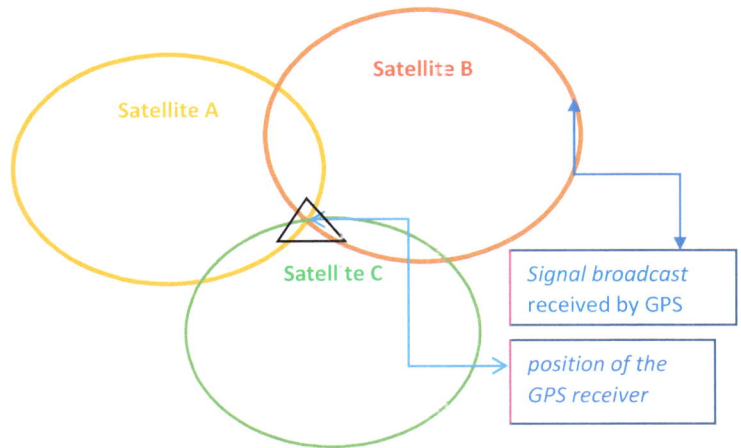

This whole process is known as **Trilateration.** To ensure the accuracy of the locations, the time signals of all the satellites must be as precise as possible. The more precise they are, the more precise the GPS receiver can calculate the distance to each of the satellites – from which it gets the current location. Now you see how vital the atomic clocks in the satellites are.

※ **Note: Triangulation is different from Trilateration.**

Special Relativity comes in during the signal sensing by the GPS ground receiver. Since the satellites are in relative motion to the ground receiver, the atomic clock signal from the satellites would be observed to be ticking slower. In a day, the clocks would lose 7 microseconds. Over time, this would mean several hours lost. What will these lead to? Inaccuracies in the position generated by the GPS Of course. It will start from little errors in the distance, and before long, you might end up spending 5 hours on a journey your car GPS says will be 5 minutes. How fun will that be!

The GPS compensates for this by accounting for Special Relativity and Einstein's second theory, which I will talk about in my next book.

# CHAPTER 4

## 4.0 Speed of Light: Can we achieve it?

The fastest car in the world right now, as at October 2019, is the Bugatti Chiron Super Sport 300+. It clocks about 304 miles per hour; 490.484 kilometers per hour. To put meaning to this figure – the car covers 450 feet in a second. That's mind-blowing! The record was set on August 2$^{nd}$, 2019, on the Volkswagen test track in Germany. On the other hand, the fastest aircraft is the North American X-15 with a speed of 4473.873 miles per hour; 7,200 kilometers per hour. That's more than six times the speed of sound, commonly called Mach 1.

All these speeds are still a far cry from the speed of light. The speed of light in kilometer per hour is about $1079 \times 10^6$. That's $7.24 \times 10^{17}$ Miles per hour. Even the escape velocity needed by space aircraft to "escape" the gravitational pull of the Earth's surface is still very small compared to it; 40,000 kilometers per hour or 25,000 miles per hour.

Can this speed of light be achieved then, not minding the huge gap that still must be filled? Special Relativity has all the answers in its delicate equations.

## 4.1 Speed of Light: Einstein's Thought

A consequence of Special Relativity theory is the idea that the speed of light is the limit of the universe. Meaning, nobody can move faster than this speed. Why? In our everyday life, we know that momentum is equal to mass multiplied by velocity.

$p = m \times v$

where p denotes momentum, m is mass, and v is velocity.

The more the mass or velocity of a body, the more the momentum – more inertia energy. The more energy is needed to change its state of motion. This required energy is its momentum. John Wallis, in 1670, had already discovered this as written in his famous work, Mechanica sive De Motu, Tractatus Geometricus.

This principle was what Einstein took further. Since momentum is conserved, then it means that, at speed closer to the speed of light, the mass of such a body would be infinite, since the momentum will be very large; and speed of light is a constant. To simplify, for a significant momentum to be obtained at the speed of light since the speed of light is constant and can't be increased, then the mass has to increase. This was a little tweak that had to be made to the equations of motion at the speed of light, to make them balanced. So far, this hasn't been disproved.

Now, I feel it must be made clear that, when we say the mass of a body increases with speed, we don't mean the mass generally taught in most basic physics classes. Mass is constant everywhere in the world. It is the quantity of matter (atoms, particles) that makes a body. A man of 50kg will be of the same mass, even on the moon. Weight, on the other hand, has to do with gravity. The weight of the same man on the surface of the earth will be different from that on the moon – why? – Because the Earth's gravitational force is much stronger than that of the moon's. Mass is, therefore, different from weight. What you measure on the scale is your mass, not weight, although people call it weight.

Apologies for the digression. When there's an increase in speed, the mass of the body increases also. This becomes only noticeable when the speed is relativistic (close to the speed of light). This doesn't mean the number of atoms and particles in a body increases, NO! So, what does it mean?

There are still a lot of arguments about what this means. Suffice to say, Einstein came up with this famous equation, called the **Energy Equivalence**.

$E = MC^2$

It's this same principle that is applied in atomic bombs. It converts the mass of atoms into energy, which is the explosion seen. Generally, we can say, energy is equivalent to mass. So, as the speed of a body increases, the mass (referred to as relativistic mass of a body) also increases. When the speed of the body reaches the limit (speed of light), the mass of that body would have gotten so massive that it would require an infinite amount of energy to move the body. What then happens when the body speed reduces? It's mass reverts to its rest mass.

Now you can see how hard it will be, getting to the speed of light. Light, its self, is made up of small particles called Photons. But these particles have Zero rest mass. This is the only reason they can attain the speed of light. Any other body with a non-zero rest mass can never achieve this speed. At least, from what we know now. I will be writing more on this soon.

# CHAPTER 5

**What Next?**

After explaining the theory of Special Relativity, Einstein began to ponder about incorporating acceleration into this beautiful phenomenon. To understand this, we must differentiate between **acceleration** and **speed**. Einstein's first theory, which is the Special Relativity, applies only to bodies in relative uniform speed to one another. This means, their movement is parallel, and the speed is not changing. This is why the theory is called "Special" because it applies to special cases of motion only. The speed is constant. It could be increased by a constant value every minute or seconds, but no directional change is involved. On the other hand, acceleration is a result of non-uniform changing speed and/or change in direction. The theory of Special Relativity doesn't apply to this form of motion.

Unifying acceleration with the theory of relativity, to form a sort of General Relativity theory became the next goal of Einstein. This he successfully did in 1915 — more about this in my next book. Thanks!

www.ingramcontent.com/pod-product-compliance
Lightning Source LLC
Chambersburg PA
CBHW040333220526
45473CB00009B/2672